TOO CUTE! BABY ANIMALS

BABY OCEAN ANIMALS

RACHAEL MORLOCK

PowerKiDS press

Published in 2025 by The Rosen Publishing Group, Inc.
2544 Clinton Street, Buffalo, NY 14224

First Edition

Editor: Rachael Morlock
Book Design: Michael Flynn

Photo Credits: Cover, p. 1 EduardStebner/Shutterstock.com; (series background) Viktoriia LiSa/Shutterstock.com; p. 5 Pole 2 Pole Images/Shutterstock.com; p. 7 vkilikov/Shutterstock.com; p. 9 Imagine Earth Photography/Shutterstock.com; p. 11 Manuel Balesteri/Shutterstock.com; p. 13 KPegg/Shutterstock.com; p. 15 Roger ARPS BPE1 CPAGB/Shutterstock.com; p. 17 Edgar Feliz/Shutterstock.com; p. 19 foryouinf/Shutterstock.com; p. 21 Christian Wittmann/Shutterstock.com.

Library of Congress Cataloging-in-Publication Data

Names: Morlock, Rachael, author.
Title: Baby ocean animals / Rachael Morlock.
Description: [Buffalo, New York] : PowerKids Press, [2025] | Series: Too cute! Baby animals | Includes bibliographical references and index.
Identifiers: LCCN 2024006199 (print) | LCCN 2024006200 (ebook) | ISBN 9781499445268 (library binding) | ISBN 9781499445251 (paperback) | ISBN 9781499445275 (ebook)
Subjects: LCSH: Marine animals–Juvenile literature. | Marine animals–Infancy–Juvenile literature. | Marine animals–Habitat–Juvenile literature.
Classification: LCC QL122.2 .M684 2025 (print) | LCC QL122.2 (ebook) | DDC 591.77–dc23/eng/20240310
LC record available at https://lccn.loc.gov/2024006199
LC ebook record available at https://lccn.loc.gov/2024006200

Manufactured in the United States of America

CONTENTS

Marine Life

Animals that are born in and around the world's oceans are called **marine** animals. Many of the cutest are marine **mammals** that spend most or all of their time in the water. But keep an eye out for cute seabirds and marine **reptiles** too!

Dolphin Calves

Bottlenose dolphins are mammals that live in warm ocean water. They're extra cute because they look like they're smiling! A mother dolphin gives birth to one calf that comes out tail first. The mother feeds her calf milk and teaches it to hunt and stay safe.

Whale Calves

Humpback whales live in all the oceans of the world. A mother humpback gives birth to a calf that's 10 to 15 feet (3 to 4.6 m) long. It weighs as much as 1 ton (907 kg)! The calf stays with its mother and drinks her milk for one year.

Sea Otter Pups

Sea otters live along Pacific Ocean coasts. Their pups are born in the water! A mother floats on her back and hugs her baby to her stomach. The pup can drink her milk this way. Pups have thick, soft fur that keeps them warm and helps them float.

Sea Lion Pups

California sea lions live in parts of the Pacific Ocean. In the summer, a mother gives birth to one pup on a beach. Newborn pups have a soft, dark-brown coat. When pups are a few weeks old, they learn to swim and hunt with their mother.

Penguin Chicks

Emperor penguins are born in **Antarctica** on the ice. A mother penguin lays an egg and a father keeps it warm until it **hatches**. Both parents take care of the chick. The chick stands on its parents' feet instead of the cold ice. Fuzzy **down** also keeps it warm.

Tern Chicks

Least tern chicks hatch on beaches along the Atlantic and Pacific Oceans. Their parents **incubate** eggs in a nest on the ground. After they hatch, the fluffy chicks stay with their mother and father for several months. Their parents feed them small fish, **crustaceans**, and insects.

Turtle Hatchlings

Loggerhead sea turtles live in the Atlantic, Pacific, and Indian Oceans. The mother lays eggs on the beach where she herself hatched! After two months in the nest, baby sea turtles hatch. The hatchlings make their way to the ocean and eat small sea animals.

An Ocean Home

What could be more adorable than a fuzzy penguin chick, a tiny turtle hatchling, or a smiling dolphin calf? Whether they're born on land or in water, baby ocean animals are cute as can be. As they grow, they learn how to make the ocean their home.

GLOSSARY

Antarctica: A continent at the South Pole.

crustacean: An animal that has several pairs of legs and a body made up of sections that are covered in a hard outer shell.

down: A covering of soft fluffy feathers.

hatch: To break open or come out of.

incubate: To keep eggs warm so they can hatch.

mammal: Any warm-blooded animal whose babies drink milk and whose body is covered with hair or fur.

marine: Having to do with the sea.

reptile: A cold-blooded animal with thin, dry pieces of skin called scales.

FOR MORE INFORMATION

BOOKS

Colón, Erica L. *Ocean Animals and Their Ecosystems: A Nature Reference Book for Kids.* Emeryville, CA: Rockridge Press, 2020.

Horning, Nicole. *Baby Penguins*. New York, NY: Cavendish Square Publishing, 2021.

Szymanski, Jennifer. *Whales*. Washington, D.C.: National Geographic, 2020.

WEBSITES

Bottlenose Dolphins
kids.nationalgeographic.com/animals/mammals/facts/bottlenose-dolphin
Learn more about the lives of friendly bottlenose dolphins.

Loggerhead Sea Turtle
www.nwf.org/Educational-Resources/Wildlife-Guide/Reptiles/Sea-Turtles/Loggerhead-Sea-Turtle#
Find out what makes loggerhead sea turtles such special ocean animals.

Publisher's note to educators and parents: Our editors have carefully reviewed these websites to ensure that they are suitable for students. Many websites change frequently, however, and we cannot guarantee that a site's future contents will continue to meet our high standards of quality and educational value. Be advised that students should be closely supervised whenever they access the internet.

INDEX